Understanding the Elements of the Periodic Table™

LITHIUM

Paula Johanson

rosen publishing's rosen central®

New York

For Scotty, who kept the Enterprise's *dilithium crystals working—and James Doohan, the actor who made him as real as my father's pacemaker battery*

Published in 2007 by The Rosen Publishing Group, Inc.
29 East 21st Street, New York, NY 10010

First Edition

Library of Congress Cataloging-in-Publication Data

Johanson, Paula.
Lithium / Paula Johanson.—1st ed.
p. cm.—(Understanding the elements of the periodic table)
Includes bibliographical references and index.
ISBN 13: 978-1-4042-0940-4
ISBN 10: 1-4042-0940-9 (lib. bdg.)
1. Lithium. 2. Periodic law.
I. Title. II. Series.
QD181.L5J65 2007
546'.381—dc22

2006006385

Manufactured in the United States of America

On the cover: Lithium's square on the periodic table of elements; the atomic structure of a lithium atom *(inset)*.

Contents

Introduction

Have you ever seen fireworks exploding in sprays of red sparks? Those red sparks are created by adding lithium compounds to the firework materials.

Lithium is an element few people ever see as a lump of pure metal. But lithium compounds are used in red fireworks and in many white ceramic and glassware pots for cooking. When you watch or read reports from deep-sea submarines and space probes to other planets, you can be sure that the metal lithium was an important part of those exciting journeys.

In weight, lithium is the lightest of all metals. The study of metals and their chemistry is called metallurgy. It is also the science of extracting metals from their ores, mixing metals, and creating useful objects from metals. Metallurgy began with the ancient use of copper (Cu) and gold (Au), which can be found in lumps of pure metal in streambeds in many parts of the world. It was through metallurgy that many metals such as lithium were discovered in ores, rocks, and even briny lakes. Lithium compounds are a minor part of most igneous rock around the world.

As industrial and medical uses were found for this light metal, lithium became more than an obscure ingredient in a few ceramics. And as scientists discovered ways to use lithium in their studies of nuclear physics, lithium became a key ingredient in the production of electric energy in nuclear power plants. It is also an essential part of something much more

Fireworks explode in bright sprays of different colors, depending on the elements that have been mixed into the firework materials. Lithium compounds add a red color to the hot sparks as they burn. This is one of the few ways you can actually see lithium in action. Most uses for lithium don't make this metal obvious to a casual observer.

dangerous—the nuclear weapon known as the hydrogen bomb, which explodes with the heat of a small sun.

Lithium is also present in plants and makes up a tiny but important part of our food and our bodies. Some prescription drugs containing lithium improve the lives of many people. It is possible that in the future we will better understand the role lithium plays in living things. Lithium is not only something to fear in nuclear weapons; it is also a vital part of many technological wonders. When you next see a display of fireworks, look for the red sparks as a visible sign of this useful element that is usually an invisible component of our daily tools.

Chapter One
The Element Lithium

The element lithium is an important part of many inventions of modern technology. Lithium's chemical symbol, Li, comes from the Greek word *lithos*, which means "stone." Lithium is never found in its pure state in nature because it reacts with air and water to form simple compounds called salts. These compounds, however, can be found in volcanic rock around the world and dissolved in the water of some lakes. In its pure form, lithium is too soft to be used to make durable goods, such as automobiles or refrigerators. But when alloyed with aluminum (Al) and other metals, lithium makes alloys strong and light enough for use in airplanes and spacecraft. Lithium is also very useful when combined with elements such as chlorine (Cl) and oxygen (O). These simple salts are light and easy to transport. Some are used because of their abilities to combine with water or gases in the air. And some are used as medicines for mental illness.

Atoms, Elements, and Compounds

Everything throughout the world—and almost everything in the entire universe—is made from elements. Some things are made of only one element, such as aluminum, copper, gold, and oxygen. Others are compounds, made of different elements combined chemically, their atoms

Atoms are made up of the subatomic particles protons, neutrons, and electrons. The protons and neutrons cling together in the nucleus. The electrons circle the nucleus in layers called shells. A drawing of an atom like this one shows the electrons close to the nucleus so the drawing will fit on the page. But a more realistic model would have tiny electrons whizzing around a stadium with a marble at the center for the nucleus.

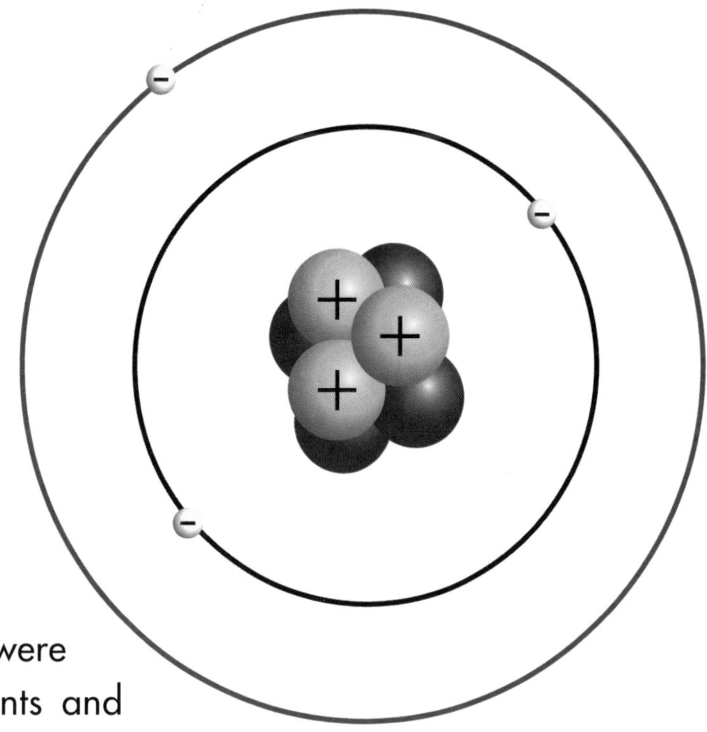

bonded together as if they were glued in different arrangements and with glues of different strengths.

Just how tiny a piece can you get of anything? A single pill of a medicine called lithium chloride looks the same all the way through when cut in half. The smallest piece you could cut would have both lithium and chloride, which are combined chemically. You could use chemical reactions to separate the lithium from the chloride. We call the smallest piece that you can get of an element an atom. It's much too tiny to see with any microscope you might use in school. Atoms of lithium are so small that about 33 million of them, side by side, would make a line only 0.4 inch (1 centimeter) long. One atom is as small a piece as you can get of lithium.

Inside the Atom

Atoms are made up of even smaller pieces called subatomic particles. There are three kinds of subatomic particles in atoms. At the atom's center, the nucleus is a dense core of protons, which have a positive electrical charge, and neutrons, which have no charge. The nucleus is very small physically—you can imagine it as a marble in the center of a model atom the size of a football stadium. Nearly all the mass of the atom is in the nucleus. Lithium has three protons in its nucleus. That is how we identify the element. Any atom with three protons in its nucleus, anywhere in the entire universe, is an atom of lithium. The nuclei of lithium atoms also contain neutrons; some have three and most have four neutrons. This is significant because the neutrons contribute as much as or more than the protons to the mass of the lithium atom.

Orbiting the nucleus of an atom, the electrons are in overlapping layers called shells. Electrons have a negative electrical charge and are attracted to the positive charge of the protons in the nucleus. The atom's negative and positive electrical charges are balanced, with an equal number of electrons and protons. There are three protons and three electrons in an atom of lithium.

In metal elements, the outer shell of electrons is less tightly bound to its nucleus than the inner electrons. We call these outer electrons valence electrons. These electrons move from one atom to the next and can be readily exchanged with neighboring atoms. This movement of electrons conducts heat and electricity through metals. An atom of metal can also give up one or more valence electrons to an atom of a nonmetal (lithium, for example, loses only one electron); then both of these atoms become ions, or charged particles. The metal becomes a cation, which means it has a positive charge because this ion is now missing one or more electrons.

The Periodic Table

Some elements, including copper, gold, iron (Fe), and mercury (Hg), have been known for thousands of years. As elements were recognized and studied over the years, scientists organized their knowledge. But a list of elements sorted by their names tells nothing about how these elements

Lithium  Snapshot

Chemical Symbol:	Li
Classification:	Alkali metal
Properties:	Silver-white color; solid at room temperature; malleable; ductile; conductor of heat and electricity; reacts quickly with air and water (which is why we never encounter lithium metal in our everyday lives)
Discovered By:	Johan August Arfwedson (also spelled Arfvedson) in 1817
Atomic Number:	3
Atomic Weight:	6.941 atomic mass units (amu)
Protons:	3
Electrons:	3
Neutrons:	3 or 4
Density at 68°F (20°C):	0.534 grams per cubic centimeter (g/cm^3)
Melting Point:	357°F (181°C)
Boiling Point:	2,457°F (1,342°C)
Commonly Found:	Volcanic rocks in Earth's crust

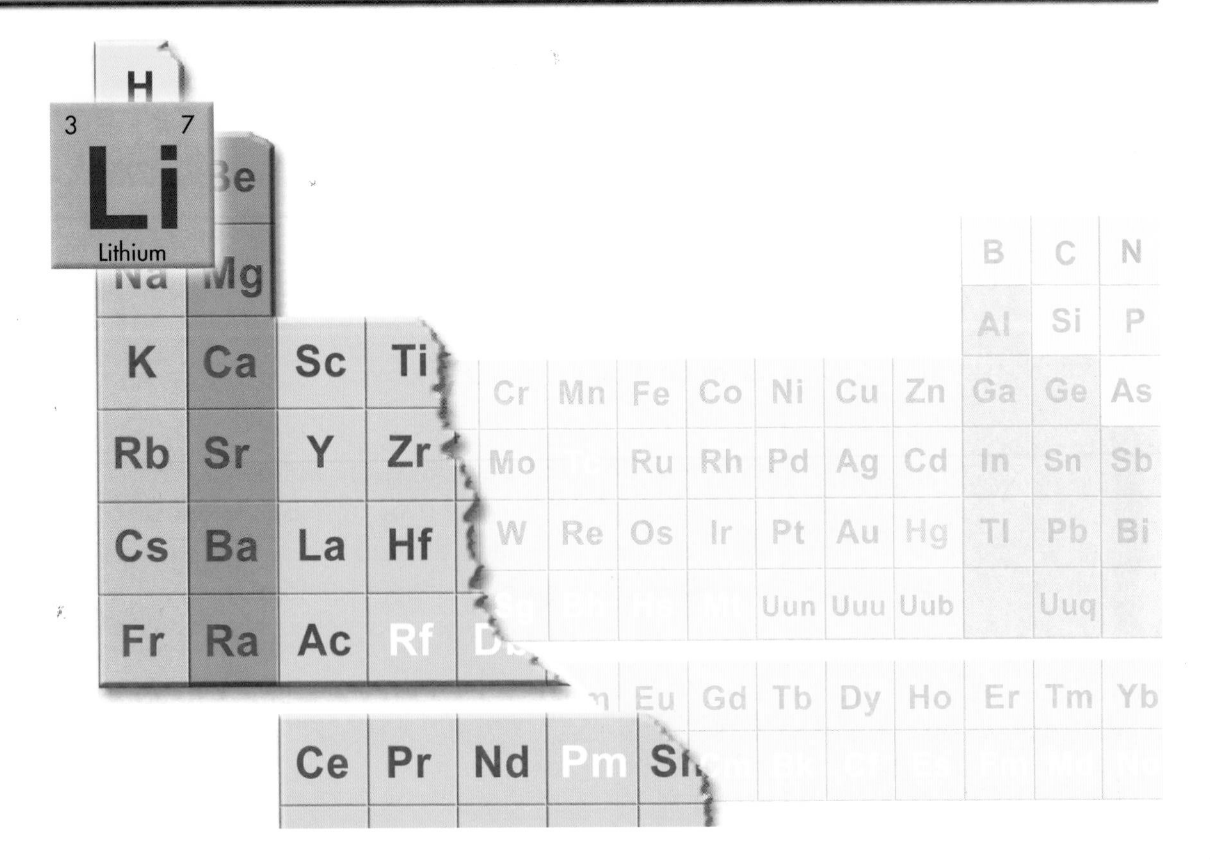

The periodic table has spaces on it for every element. Each element is shown in a box with its name and its chemical symbol, which is a short version of its name (usually in English or Latin). On the left is written the number of protons in an atom of the element, called the atomic number. For lithium, this is 3. On the right is written the total average number of protons and neutrons in an atom of the element, called the atomic weight or atomic mass. For lithium, this is rounded to 7.

are different or alike. Russian chemist Dmitry Mendeleyev (also spelled Mendeleev) created a chart of elements sorted by physical properties, later called a periodic table. He made his chart in 1869 at the University of St. Petersburg in Russia to teach his students. Mendeleyev used his chart to make predictions about elements, many of which were later confirmed. He was the sixth person to publish a chart of the elements, and his version became the one most widely known. Although today's periodic table has been revised and expanded since Mendeleyev created his chart, the basic look is the same. (See the periodic table on pages 42–43.)

The periodic table organizes elements by increasing atomic number (how many protons are in the nucleus of an atom of the element), and also by trends in reactivity among the elements. (Reactivity describes how easily an element interacts with other substances.) The periodic table's most apparent distinction is the difference between metals (in the bottom and left parts of the chart) and nonmetals (in the upper and right parts of the chart). Nonmetals and metals are separated by a steplike line across the table. Metals are located to the left of this line.

Most metals share certain qualities. They conduct electricity (some better than others). Metals can be made shiny by polishing. Most metals are malleable, or can be beaten into shapes without breaking; these malleable metals are usually also ductile, which means they can be pulled into wires. Metal atoms also form ions with positive electrical charges (called cations) when they give up one or more electrons from their outer electron shells; although this loss of electrons itself cannot be observed, the effects of the electron loss can be. Metals react with nonmetals to form compounds with high melting points.

Grouped together on the right of the steplike line are the nonmetals. On the far right of the chart are the noble gases, which do not react with any elements under most conditions. The other nonmetals are able to form ions with negative electrical charges (called anions) by picking up one or more extra electrons. Metalloids, such as boron (B) and silicon (Si), are located on the immediate right of that steplike line, between metals and nonmetals. In most respects metalloids behave like nonmetals. They do, however, conduct electricity, though less well than metals.

Lithium is located left of the line, among the alkali metals—lithium, sodium (Na), potassium (K), rubidium (Rb), cesium (Cs), and francium (Fr)—in the leftmost column of elements. Alkali metals are more reactive than the transition metals that may be more familiar to you, such as iron or copper. They also have similar characteristics: they are malleable and ductile, and they conduct heat.

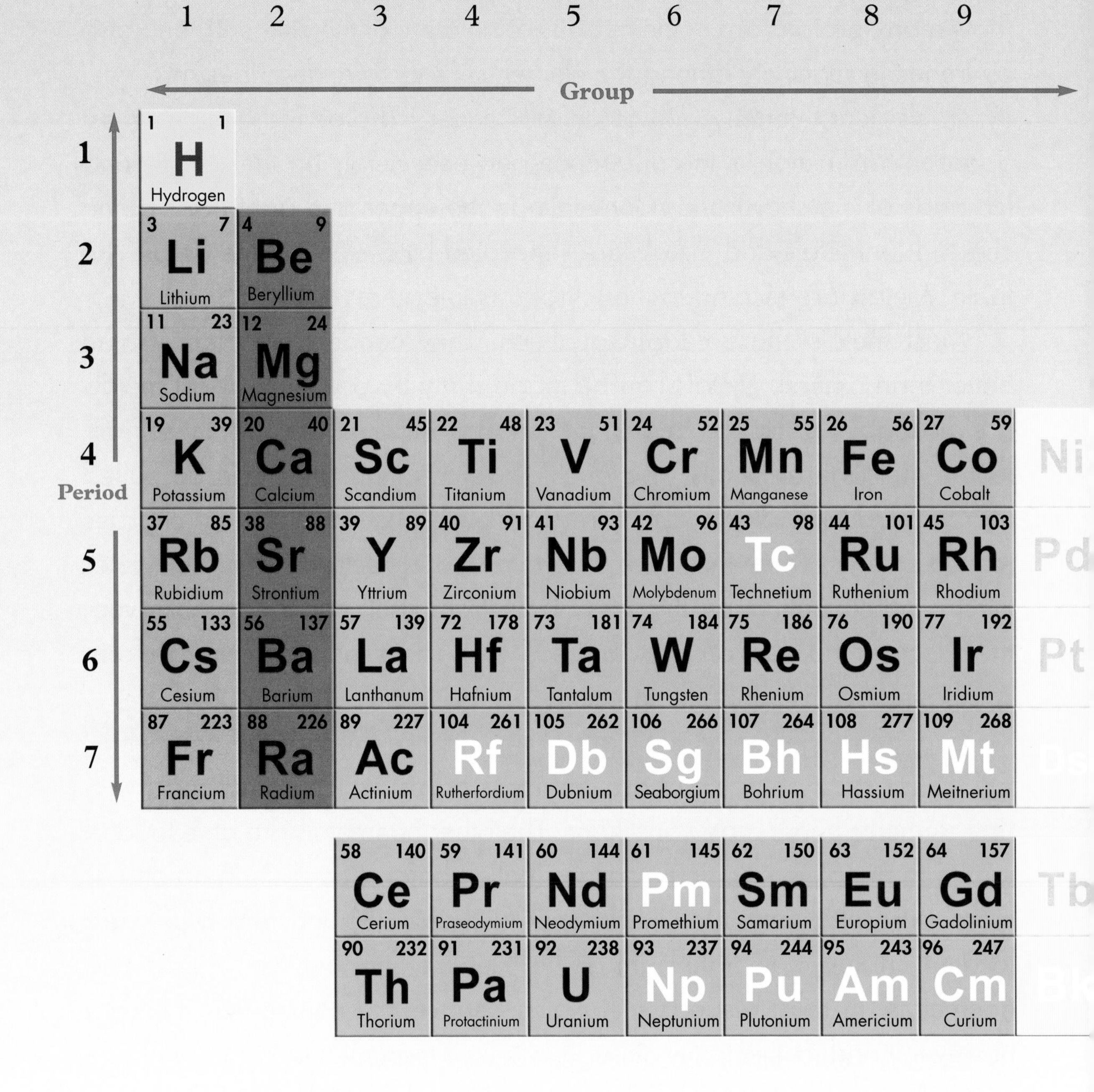

The periodic table sorts elements into groups and periods. Each element in a group has similar properties. Each element in a period has one more proton than the element on its left, and one fewer than the element on its right; this affects its physical and chemical properties. Studying these properties helps us understand the relationships among the elements. The steplike line, which can be seen in the complete table on pages 42–43, divides metals from nonmetals. Metalloids are on the immediate right of this line. Lithium is on the far left, among the metals.

Periods and Groups

Each horizontal row of elements in the periodic table is called a period. Elements in one period all have the same number of electron shells surrounding the nucleus of each atom. Because lithium has two shells of electrons, it is in period 2. All the elements in period 2 are in the second row down from the top of the chart. Each element in the row has one more proton in its nucleus than the element on its left, and one more electron in its outer shell.

A vertical column of elements is a group. Elements in a group have similar properties. Lithium is in group IA (also called group 1) with the alkali metals. All the elements in group IA are in the first column on the left side of the chart. Lithium has two electron shells, sodium has three, and potassium has four. Group IA elements are able to give up one electron easily, but they combine with oxygen more quickly than group IB elements (copper, silver, and gold). Group IIA elements give up two electrons, and group IIIA elements lose three electrons. Group IA elements are more likely to form simple compounds, while group IB elements are more likely to form complex molecules.

Unique Qualities of Elements

We define an element by the number of protons found in the nuclei of its atoms. Any atom with three protons in its nucleus is an atom of lithium. In the periodic table, the atomic number 3 is written above and left of the symbol for lithium (Li). An atom with four protons would be another element, beryllium (Be), with different properties. An atom with two protons would be another element, helium (He). The presence of one proton more or fewer makes a big difference: beryllium is a hard metal, lithium is as soft as cool butter, and helium is a gas except at extremely low temperatures. Atomic numbers also indicate the number of electrons, unless the atom is an ion.

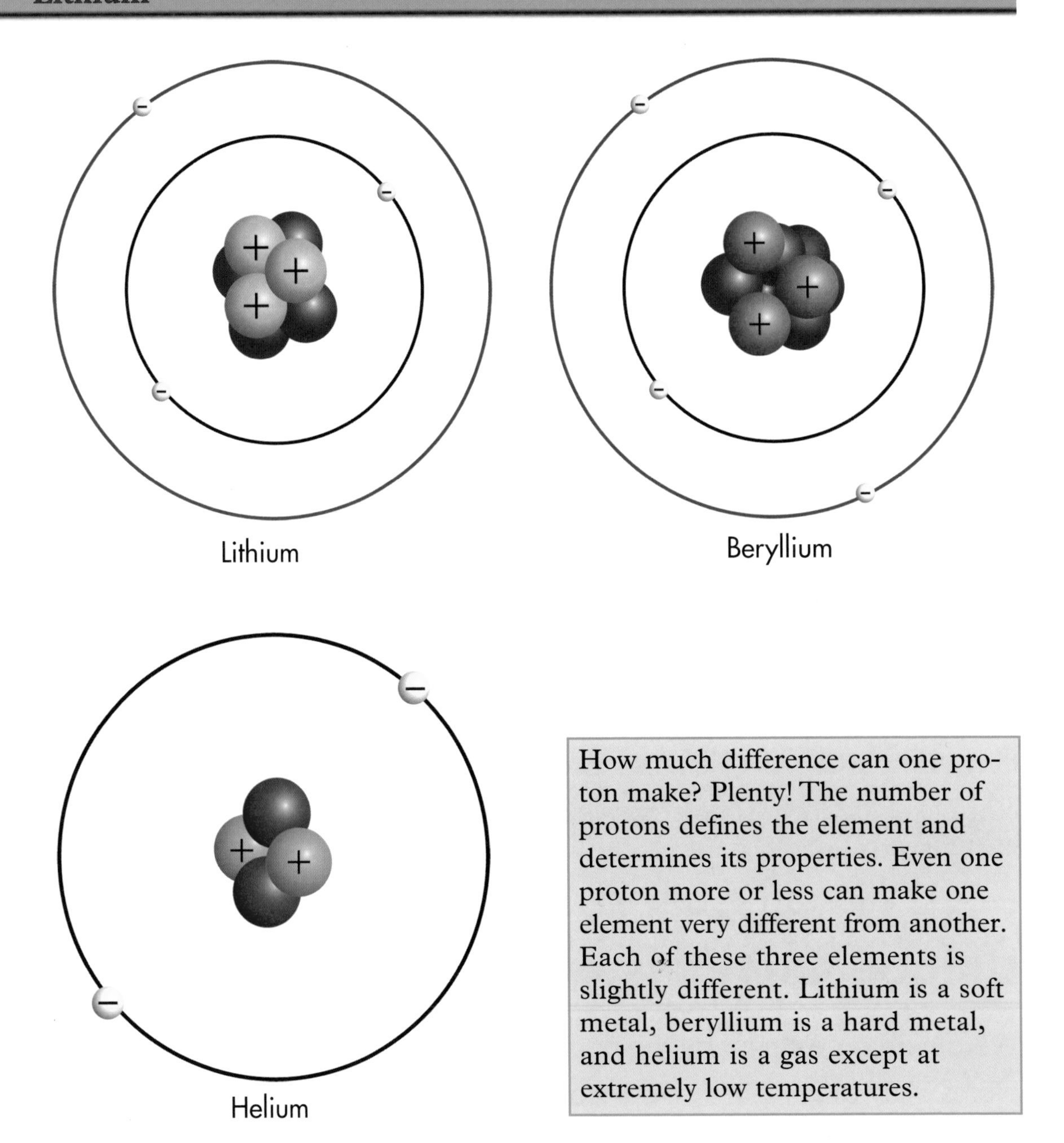

How much difference can one proton make? Plenty! The number of protons defines the element and determines its properties. Even one proton more or less can make one element very different from another. Each of these three elements is slightly different. Lithium is a soft metal, beryllium is a hard metal, and helium is a gas except at extremely low temperatures.

Atomic Weight

In the periodic table, the number 7 is written above and to the right of the symbol for lithium. That is the atomic weight of lithium—the number of protons plus the average number of neutrons found in an atom of that element. Atomic weight is also called atomic mass. It is measured in atomic mass units (amu). The atomic weight of lithium is 6.941 amu, which has been rounded to one digit, 7, in our periodic table.

Atomic weight is an average weight, not exact, because atoms of an element are not always identical. Some lithium atoms (92 percent) have four neutrons, while others (8 percent) have three neutrons in the nucleus. The lithium atom with four neutrons has more weight or mass than the one with three neutrons, but this does not affect the element's physical or chemical properties, which are determined mostly by the electrons in the atoms. The nuclei of many elements have two or more possible numbers of neutrons. We call these alternate versions isotopes.

Chapter Two
The Properties of Lithium

Elements are described by chemical and physical properties. Scientists use these characteristics to identify and classify an element. The physical properties of a pure element, such as hardness, density, and conductivity, can be observed. We can also observe the element's phase at different temperatures. At room temperature, is it a solid? Does it melt into a liquid at a very low or very high temperature? What is the boiling point?

We mix atoms of a pure element with other substances to observe how it reacts (called reactivity). These chemical properties describe the element's abilities to combine with other elements into new molecules. One chemical property of lithium is its ability to combine with oxygen to form oxides. This happens very easily. If you expose a piece of pure lithium to air, the surface will oxidize within a minute. Some other metals oxidize more slowly (such as iron rusting), or only when heated to high temperatures (like copper). During a chemical reaction, the element is changed from a pure substance containing only one kind of atom to a new substance, a compound of two or more kinds of atoms chemically bonded together. These ions fit together as if the atoms were glued in various ways. For example, when lithium quickly reacts with oxygen in the air, it changes from a soft, silvery, and shiny metal to form a dull gray coating of tarnish.

A piece of pure lithium is soft enough to cut easily with a knife. The freshly exposed edges are shiny and bright, but in only a few minutes the surface of the metal combines with oxygen from the air to make lithium oxide. This dulls the surface of the metal to gray. Pure lithium is stored in oil, to keep it from being exposed to air.

Lithium's Phase at Room Temperature

At room temperature, elements are gas, liquid, or solid. Knowing the physical state or phase of an element at room temperature helps scientists in their experiments with chemical reactions. Most elements are solids at room temperature, and all metals are solid except mercury. At room temperature, lithium is a shiny, silvery solid. A solid keeps its volume and shape, and under pressure or compression it resists changing that shape.

Lithium's Density

Density measures how much mass an object contains in a certain volume (mass per unit volume, frequently expressed as grams per cubic centimeter). Lithium's density is 0.534 g/cm^3. Liquids are denser than gases, and solids are usually denser than liquids. However, if you drop a solid piece of lithium into water, it will float because lithium is less dense than water. (It will also react with the water, releasing gases and heat, which could be dangerous!)

Lithium's Hardness

Pure lithium is a very soft metal, too soft to be useful for tools. To make it harder, other elements can be added. When lithium and magnesium are combined, these two soft and very light metals make an alloy that is very light and strong, especially when compared with iron or copper.

The hardness of common substances is measured with Mohs' scale. Friedrich Mohs, a German mineralogist, learned from miners who routinely used scratch tests to compare minerals. He published a scale in 1822, comparing ten groups of substances, ranked in order of increasing hardness. Substances from 1 to 2.5 on his scale are considered soft. Those of an intermediate hardness are between 2.5 and 5.5, and substances at 5.5 and above are considered hard. Diamond is the hardest substance of all, in the tenth group. On Mohs' scale, lithium has a hardness of 0.6. A piece of talc rock has a hardness of 1, and your fingernail has a hardness of 2.5. Your fingernail could scratch a talc rock, but talc could scratch the lithium. You would never scratch lithium with your fingernails. It would react with your skin and could cause a corrosive burn.

Mohs' Scale

Hardness Rating Examples	
0.6	Lithium
1	Talc
2	Gypsum (and rock salt, fingernails)
3	Calcite (copper)
4	Fluorite (and iron)
5	Apatite (and cobalt [Co])
6	Orthoclase (and rhodium [Rh], silicon, tungsten [W])
7	Quartz
8	Topaz (and chromium [Cr], hardened steel)
9	Corundum (sapphire)
10	Diamond

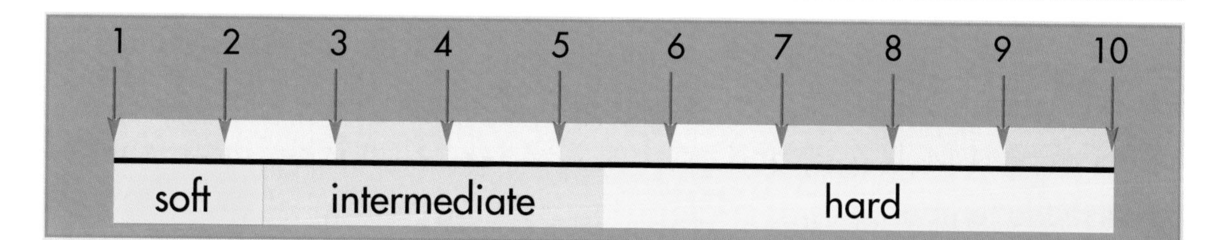

Fire Hazard

If pure lithium catches fire, it burns very easily with a bright, white flame that can scorch even asbestos. Ordinary fire extinguishers would not be able to put it out. Only Class D extinguishers would work. These spray copper powder, the only known agent that clings to a vertical or flowing surface and smothers burning liquid lithium.

Lithium: Conductor of Electricity and Heat

Lithium conducts electricity and heat, as all metals do. Valence electrons in the outer shells of metal atoms can move from one atom to another. We call this movement or flow of electrons electricity. This movable sea of electrons also makes metals good conductors of heat. The heated electrons move about quickly, and so heat is distributed throughout the piece of metal.

There is only one valence electron in each lithium atom, so the energies binding lithium atoms together in a metallic lattice are relatively weak compared to some other metals such as iron. Because of this, pure lithium is a metal that melts at a low temperature, and when solid it is as soft as chilled butter.

Lithium has the highest specific heat capacity of any solid element. Specific heat measures how much energy must be added to a substance to increase its temperature. It also measures how well a substance stores heat. Lithium's specific heat is 3,582 joules per kilogram kelvin (J/kg·K), the amount of energy required to raise the temperature of one kilogram of lithium by one degree Celsius or Kelvin. For comparison, iron is 444 J/kg·K and gold is 129 J/kg·K. It takes more energy to raise the temperature of a kilogram of lithium than a kilogram of iron or gold.

Lithium and lithium compounds are very useful under hot conditions as a heat transfer medium, to absorb heat from another object. Some lithium compounds make good industrial greases. When metals are being welded or soldered together, lithium is used as a flux to promote the fusing. It also absorbs impurities from other metals and eliminates oxidation that could weaken the weld.

Lithium is used to remove gases from other metals when they are melted. For example, if oxygen and other gases have dissolved or are suspended in melted copper, when the copper cools to a solid, there will be

Did you ever grease a bicycle chain so the bicycle would roll easily? Many machines need grease to lubricate moving parts and reduce friction. Lithium compounds make very effective industrial greases because they're not only slippery, they can absorb heat from friction.

bubbles in it. Lithium is added to liquid copper, in a copper container, to absorb the impurities. Also, adding a little lithium to aluminum—less than 1 percent of the final mix—will make the alloy harder and better able to absorb heat and to fuse.

Chapter Three
Where Can Lithium Be Found?

Lithium is the thirty-first most abundant element in the world. There is a small amount in meteorites that fall to Earth from space, too. Scientists know, from studying the light of stars and the reflected light from planets and asteroids, that there is lithium elsewhere in the universe. For every billion atoms in the universe, one is lithium. Most atoms are hydrogen and helium in big clouds of gas and in stars. Almost all the lithium in the universe is found in the rock of planets and asteroids.

Discovery of Lithium

Near Stockholm, Sweden, is the island of Utö, where José Bonifácio de Andrada e Silva, a visiting Brazilian scientist, found an interesting rock in 1800. The rock contained the minerals petalite and spodumene. Several chemists noted that petalite gives a red color to flames, but they did not investigate why. Johan August Arfwedson was a student of Jakob Berzelius's, and in 1817 he did a routine analysis of the composition of petalite ore from a mine on Utö. The ore was $LiAl(Si_2O_5)_2$: roughly 80 percent silica (like beach sand), 17 percent alumina, and 3 percent alkali. This alkali did not react like potassium or magnesium. He wondered if the alkali might be a compound of sodium, but this alkali was 5 percent larger than the size of the same mass of a compound of sodium. Arfwedson carefully repeated

The gas in a Bunsen burner normally burns with a colorless or slightly yellow flame. But when a piece of lithium carbonate is burned, it turns the flame red.

his experiments twice before concluding this was a new element, with a greater capacity to react than the other alkalis. His mentor, Berzelius, spread the word about the discovery. They named this alkali "lithium" for the Greek word *lithos*, meaning "stone," because it was derived from a mineral. The other two common group 1 elements, sodium and potassium, were discovered from plant sources.

In 1818, Christian Gmelin, a German chemist, was the first to observe formally that lithium salts give a bright red color to a flame, unlike pure lithium, which burns white. Neither he nor Arfwedson were able to isolate lithium from its salts. Pure lithium was first isolated in 1818 by William Thomas Brande and Sir Humphrey Davy, who ran an electric current through melted lithium oxide. Lithium metal wasn't available in commercial quantities until 1923, when a German company, Metallgesellschaft AG, used electrolysis on a mixture of lithium chloride and potassium chloride.

Group 1 Metals

Lithium is the lightest solid element and the lightest element in group 1, the alkali metals. These metals are so chemically active, you would never find in nature a piece of pure lithium, sodium, potassium, rubidium, or

Star Trek's Dilithium

A fictional version of lithium called dilithium crystals figures prominently in *Star Trek*. Through this popular television show's five series, ten films, and an animated series, the starship *Enterprise* speeds from one star system to another using engines powered somehow by dilithium crystals. Perhaps these rare, lemon-sized crystals were supposed to be an isotope of lithium or a compound with two lithium atoms. The only time dilithium was shown on *Star Trek*'s first series (televised from 1966 to 1969), the crystals looked much like petalite. Was this a happy accident, or did the show's prop designer know how lithium was discovered?

cesium. These elements are always combined into stable salts in minerals, and they were found and named by chemists only after experiments to discover new elements in plants and rocks.

These aren't usually the first metals that come to mind. Gold and copper are found in flakes and nuggets lying in streambeds in many parts of the world. Even iron and tin can be recognized in ores. Who would look for a useful metal in volcanic rock? But there is a little lithium in most igneous rocks. Solid deposits of spodumene, a greenish to pinkish or lilac mineral, are found in North Carolina, South Dakota, and Quebec.

While pure lithium is not found in nature, its salts are found throughout the world in igneous rock. From this source, there is lithium dissolved in seawater, in mineral-rich springs, and in briny lakes. There are briny lakes in Nevada and California, but most of the commercially sold lithium comes from briny lakes high in the mountains of Chile.

In seawater, for every billion molecules of water there are usually 140 to 250 dissolved lithium ions. But a sample of water from near a hydrothermal vent might be different. These vents are places deep in the ocean where water heated by underground magma is released. There are more minerals than usual dissolved in the hot water coming from these vents, and lithium can be present in nearly 7,000 parts per billion.

Lithium Used to Prove Einstein's Theory

Physicist Albert Einstein's 1905 equation $E=mc^2$ was the result of a thought experiment. Einstein believed mass and energy were the same thing. He figured you'd have to multiply the mass of an object by the speed of light (that's 186,000 miles [299,338 kilometers] per second) squared, to know

High in the Andes mountains is this lake with no outlet. It has become briny (salty) from evaporation. There are a lot of mineral salts in this briny lake, dissolved out of igneous rock, including lithium salts. This lake and other briny lakes are good industrial sources for lithium, instead of trying to mine the lithium out of volcanic rock.

the object's energy. If you could convert even a small mass into energy, it would be far more energy than could be released by chemical reactions. For people used to chemical reactions like burning about two pounds (one kilogram) of coal to heat a stove, it was hard to imagine that the same two pounds of coal was the equivalent of enough energy to heat a billion stoves. But how could you turn that coal into energy? Einstein suggested that radioactivity be studied to see if the binding energy that holds the different particles inside the nucleus of an atom could be suddenly released.

Lithium's Atom

When you use a cue ball to hit billiard balls on a pool table, the other balls don't get any more energy than the push you gave the cue ball. So why does a proton hitting an atom of lithium release so much energy when the atom breaks into two ions of helium? It's because an atom isn't made of balls sitting side by side on a table. The nucleus has protons and neutrons held together by terrifically strong forces. When that nucleus is struck by a fast-moving proton, it becomes unstable and will break into more stable parts. That breaking can turn a small amount of mass into energy.

It would be harder to play pool if the balls were held together with rubber bands that might suddenly snap and fling the balls apart, or even break them! The energy that bonds protons and neutrons together in a nucleus is very strong. When that bond is broken, the protons and neutrons can be flung around by the released energy.

The presence of lithium in lithium carbonate (Li_2CO_3) can be observed when a small amount of the compound is heated. In photo 1, the fireproof crucible on the left contains ground-up lithium carbonate, and the one on the right, a Tums tablet, which includes some calcium carbonate. In photo 2, after ethanol had been added to each, the mixture was lit and the lithium carbonate burned with a red flame and the Tums burned with a yellow flame.

In 1932, J. D. Cockcroft and E. T. S. Walton at the Cavendish Laboratory in Cambridge, England, succeeded in turning a tiny bit of mass into energy. They accelerated hydrogen ions (protons) aimed at a target of lithium metal. Each lithium atom that was hit turned into two helium ions with many times more energy than the "proton bullets." Where did the extra energy come from? The atomic mass of an atom of lithium plus a proton is just 0.02 amu more than two ions of helium, and that mass times the speed of light squared equals the extra energy—just as Einstein predicted.

Chapter Four
Lithium Compounds

For decades after its discovery in 1817, lithium was only a laboratory curiosity. Most properties and uses of pure lithium and its compounds were found by trial and error.

Lithium is the lightest of all solid elements. Its density is so low that a piece of pure lithium can float on water, oil, or even gasoline. But don't try it! This element is highly reactive at room temperature. It will combine with water, oxygen, carbon dioxide, and even nitrogen or other elements in reactions that release a dangerous amount of heat and gases.

Though lithium itself is very reactive, many lithium salts are stable and inert as sand. For decades, almost the only use for lithium-bearing minerals was as an exotic additive to ceramics. Ceramics with these lithium compounds have a very low thermal expansion: they do not expand much when heated, or shrink much when chilled. These ceramics have excellent thermal shock resistance, so they do not crack when suddenly heated. When the space program took an interest in these ceramics, the result was heat shields to protect astronauts returning to Earth from the moon.

Military Uses of Lithium Compounds

During the 1940s, lithium hydride (LiH) was used to generate hydrogen (H) to inflate weather balloons and emergency signaling balloons, particularly from ships at sea. One pound (0.454 kg) of lithium hydride combines with seawater to generate 45 cubic feet (1,274 liters) of hydrogen gas, which could lift a balloon and its small cargo high into the air. Lithium hydride was also used to inflate life rafts, which saved many sailors and navy personnel.

The primary use of lithium during and after the Second World War (1939–1945) was in the nuclear industry. Beginning in 1953, the U.S. government's Atomic Energy Commission was the largest consumer of lithium in the world, separating lithium-6 isotope from large amounts of lithium hydroxide to use for making thermonuclear weapons. Many of these weapons were stockpiled, but testing was limited. Only two nuclear weapons were ever used in wartime at the end of World War II.

After the Atomic Energy Commission contracts expired in 1960, American industry found civilian uses and particularly nonnuclear applications for lithium. Many technologies have come to rely on this unique element. Some uses require small or trace amounts.

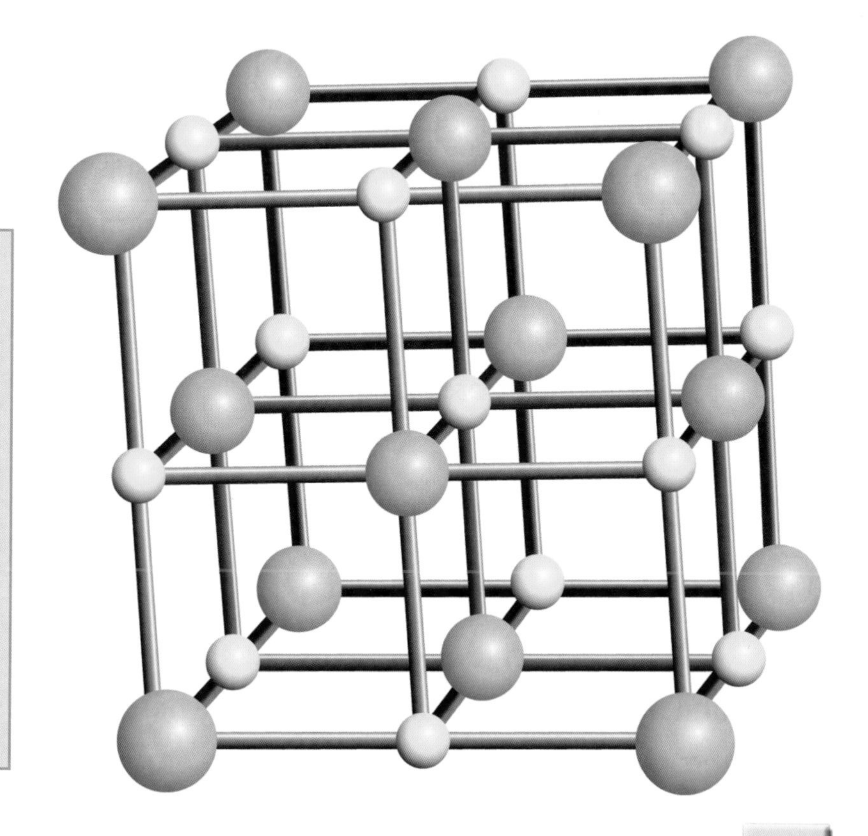

Two ions (like lithium+ and hydride-) can combine to form a compound. In lithium hydride, the ions are close to the same size, so when solid they can stack neatly into a regular lattice. When exposed to seawater, the lithium ions can combine readily with other ions in the seawater and with water molecules. Hydrogen gas would be released.

Breathable Bottled Air

How do we keep the air breathable in a submarine or a spacecraft? People exhale carbon dioxide (CO_2) with every breath. But lithium hydroxide (LiOH) is used to take carbon dioxide from the air by combining with it to form lithium carbonate and water ($2\ LiOH + CO_2 \rightarrow Li_2CO_3 + H_2O$). Any alkali hydroxide will absorb CO_2, but lithium hydroxide is the best choice when designers are trying to keep the submarine or spacecraft as light as possible.

Lithium Chloride

It's easy to dry up water drops on a table using a towel. But how do you get water molecules out of air or a beautiful flower? Lithium chloride (LiCl) is a drying agent that binds with water molecules, one of the most effective materials for this purpose. It also can dissolve in water, unlike some other lithium compounds.

The primary means for obtaining pure lithium is running an electric current through melted lithium chloride. The high temperature (1,137°F [614°C]) at which lithium chloride melts makes this process expensive, but adding sodium chloride (NaCl) lowers the melting point, making the process more affordable.

Lithium Hydroxide

Glass is so inert, it doesn't react with many substances. Though you wouldn't expect to find any lithium (which is very reactive) in glass, a little lithium hydroxide can be added when the glass is being made. Some ceramics contain lithium hydroxide, too. Less than 1 percent of the final product is lithium, but it's enough to change the way the ceramic or glass reacts to temperature changes. Some glass or ceramic casserole dishes that contain lithium can be put right into a preheated oven from a refrigerator without breaking.

Lithium in Sunglasses

Some people who wear prescription eyeglasses were delighted with the invention of lenses that darken when exposed to ultraviolet light in sunlight. The glass in these lenses contains a silver halide as a light-sensitive agent, but lithium carbonate is added to help the silver halide dissolve and to control the alkalinity of the glass.

Lithium in Cement

Roads and buildings all over North America are being weakened by alkali-silica reactivity. This reaction between cement and aggregate in concrete forms a gel that expands when it is wet and cracks concrete. But if fresh cement has lithium compounds added, the effect is reduced or eliminated. This additive will ensure that concrete structures last much longer.

Lubricants

Lithium hydroxide is a strong base. When it is heated and combined with a fat, the resulting lithium soap thickens oils and is used to make lubricating greases. (Soap is made by heating fat or vegetable oil with an alkali.)

Lithium stearate ($LiC_{18}H_{35}O_2$), a soap, is used in making some industrial greases that are effective lubricants at both very low and very high temperatures. Most lubricants work well at one or the other.

Telescopes

The glass for the 200-inch (5.08 meters) reflector of the Hale telescope at the Palomar Observatory north of San Diego, California, contains lithium—not much, but enough to make the glass stronger and resistant to temperature change. Astronomers want to be sure that telescopes will last for many years. The Hale telescope was the largest telescope in use from 1948 until 1993, when the twin Keck telescopes were made at the W. M. Keck Observatory at Mauna Kea in Hawaii.

The discovery of a grease that could be used over a wide temperature range, for a variety of purposes in extreme conditions, was important for industry. It's something like a 10W40 motor oil used in automobile and bus engines. This oil is not too thick when an engine is starting on a cool winter morning, at 0°F (–18°C), and is not too liquid when the same engine is running hot after a long run pulling a heavy load uphill, at 210°F (99°C). Lithium stearate makes greases that work over an even wider temperature range.

Stable Lattice

The lithium ion Li^+ is exceptionally small. Because of this, it has an especially high charge for its size. When lithium ions combine with anions of other elements with low atomic numbers, such as nitrogen, these compounds (called salts) are very stable. The ions are similar in size and form a lattice that binds together tightly. The salts of lithium melt at higher temperatures than salts formed with other group 1 metals. When lithium ions combine with a larger anion, such as carbonate (CO_3), the resulting salts are less stable and melt at a lower temperature.

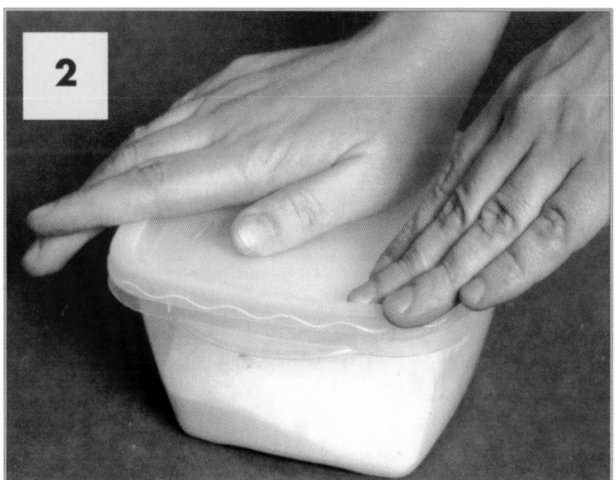

Silica gel (silicon dioxide, SiO_2) can be used as a drying agent to dry and preserve fresh flowers. The flowers' petals should be completely covered by the granules (photo 1) and kept in an airtight container (photo 2). The flowers will dry in about a week (photo 3). Some silica gel granules can include a moisture indicator, blue granules, that will turn pink when water has been absorbed.

It's harder to form a stable lattice when packing large and small ions together. The ions of other group 1 metals have a +1 charge like lithium but are much larger, with more than one electron shell surrounding their nuclei.

Chapter Five
Lithium and You

Lithium is present in many aspects of our everyday lives. Lithium compounds can be very visible in fireworks and can be handled in household objects. But there are plenty of ways that lithium affects our lives without being obvious or recognized. And there is lithium in places you might never expect to find a metal.

Lithium Batteries

Some batteries are good for only a single use. But lithium ion batteries keep their charge longer than cheaper batteries and are rechargeable. Lithium is used instead of a carbon rod in a dry cell battery. Lithium batteries also produce a higher voltage, 3.0 volts instead of 1.5 volts.

Cardiac pacemakers contain lithium batteries. These long-lasting batteries can stay in a heart patient's chest for up to ten years before needing to be changed!

Most modern cell phones use lithium batteries, and so do other small electronic devices. The telecommunications market uses lithium niobate ($LiNbO_3$) extensively for mobile phones and optical modulators for fiber-optic cables. There are lithium batteries in the Mars explorers and other space probes.

Lithium batteries can look like the batteries you put in a flashlight. Other types of lithium batteries are packaged together, and the package is put into a camera or computer. Lithium batteries for a pacemaker can last up to ten years without needing to be changed!

Lithium in the Aerospace Industry

Airplanes and helicopters are relied on worldwide for fast transport of people and goods. Modern aircraft are made from strong, light alloys of lithium with aluminum, cadmium (Cd), copper, and manganese (Mn). There's less than 3 percent of lithium in some of these alloys, which make light, high-performance parts that stand up to terrific stresses.

Lithium alloys are a vital part of the space program, too. It's important not to have dissolved gases in metal objects sent out into space, not only because gases encourage corrosion, but also because the gases will be forced out and weaken the metal when they are exposed to a vacuum. Reliable satellites operate for years in orbit. Space probes like *Galileo* and *Voyager 1* and *2* reported for years on other planets and asteroids.

Even if you've never met a pilot or an astronaut, you've probably eaten bananas and other food shipped by plane. Long-distance phone calls are carried by satellite. People in every community rely on weather

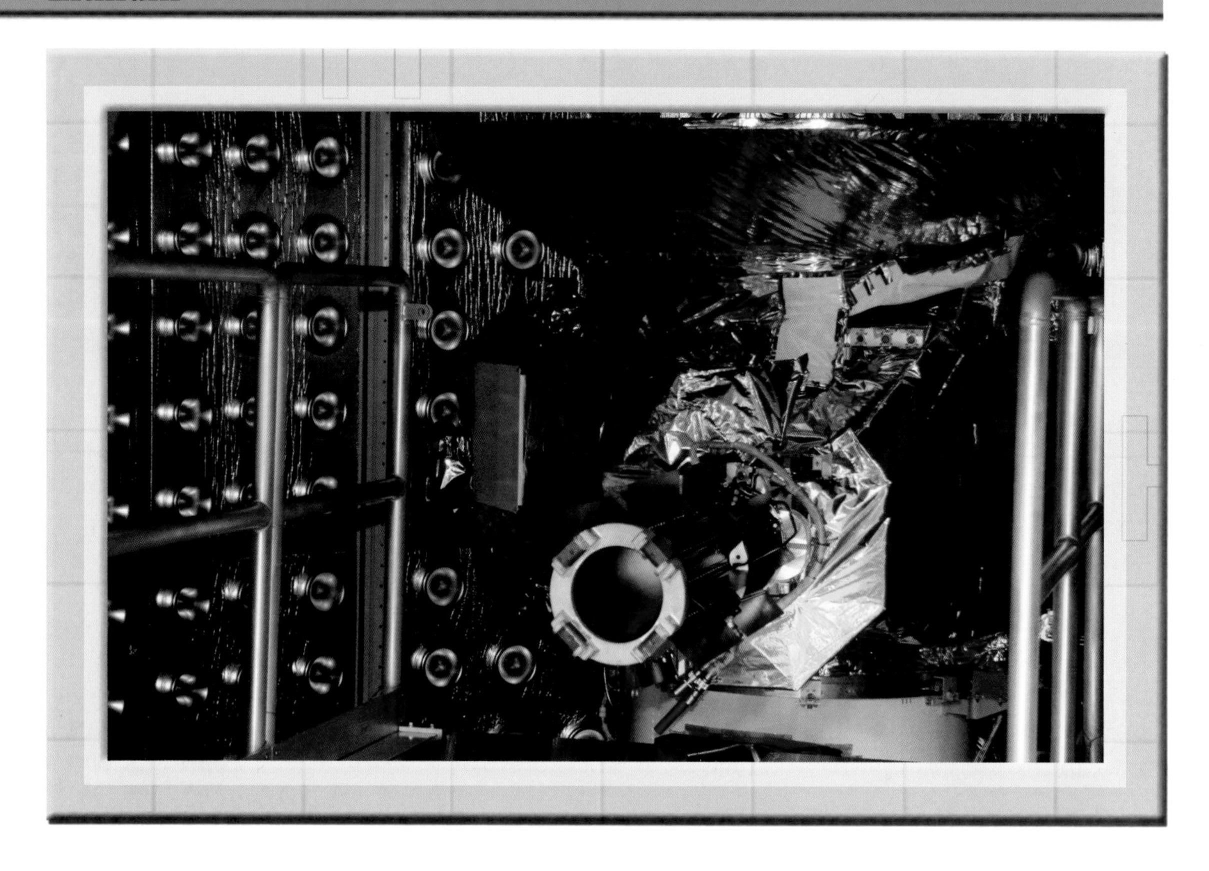

In January 2006, NASA launched the space probe *New Horizons* on a mission to the solar system's ninth planet. Lithium alloys are an essential part of the probe and also of its launch rocket. After taking photographs of Pluto and its moon Charon in 2015, *New Horizons* will go on to other distant solar system objects in the Kuiper Belt.

satellites for accurate weather reports. Some lithium alloys used in planes and space vehicles are also used in cars, trucks, and buses.

Lithium in Dehumidifiers

If you've ever been in a humid climate, you know how damp air makes a hot day unbearable. A cold, damp day can feel even colder, too. High humidity causes problems in buildings as well. Too much moisture encourages molds and mildew, some of which cause allergies or illness, such as Legionnaire's disease. Other molds and mildew ruin walls and furnishings.

Fine musical instruments and computer components are particularly vulnerable to high humidity. Lithium chloride and lithium bromide (LiBr) are used as desiccants (agents that remove moisture from the air) in drying systems, in air conditioners, and in some dehumidifiers.

Electric Power Generation

Lithium may be involved in generating electricity you use in your home and school every day. Perhaps that electricity is generated in a nuclear power plant, as is 16 percent of the electric power used around the world.

Nuclear power plants use a type of water called heavy water to absorb heat from radioactive fuel rods. A molecule of regular water has two atoms of hydrogen and one of oxygen, but in a molecule of heavy water, one or both of the hydrogen atoms is replaced with deuterium or tritium, isotopes of hydrogen. Heavy water doesn't look or taste different, but it is able to hold more heat. The harmful effects of heavy water have not been thoroughly investigated, but they seem to be minor. The only way you could tell heavy water from ordinary water by looking is to make an ice cube of heavy water. The "heavy" ice cube will sink in a glass of ordinary water, unlike an ice cube made of ordinary water, which would float.

Deuterium and tritium occur naturally, and a few molecules of heavy water can be sorted out of thousands of gallons of seawater. But the principal source of deuterium and tritium for making heavy water is lithium. If the nucleus of an atom of lithium is struck by a fast-moving neutron, it will break apart into two atoms of deuterium and one of tritium.

Hydrogen Bombs

When a hydrogen bomb explodes, the fuel for the fusion reaction is lithium deuteride (LiD), a compound of lithium and deuterium. When

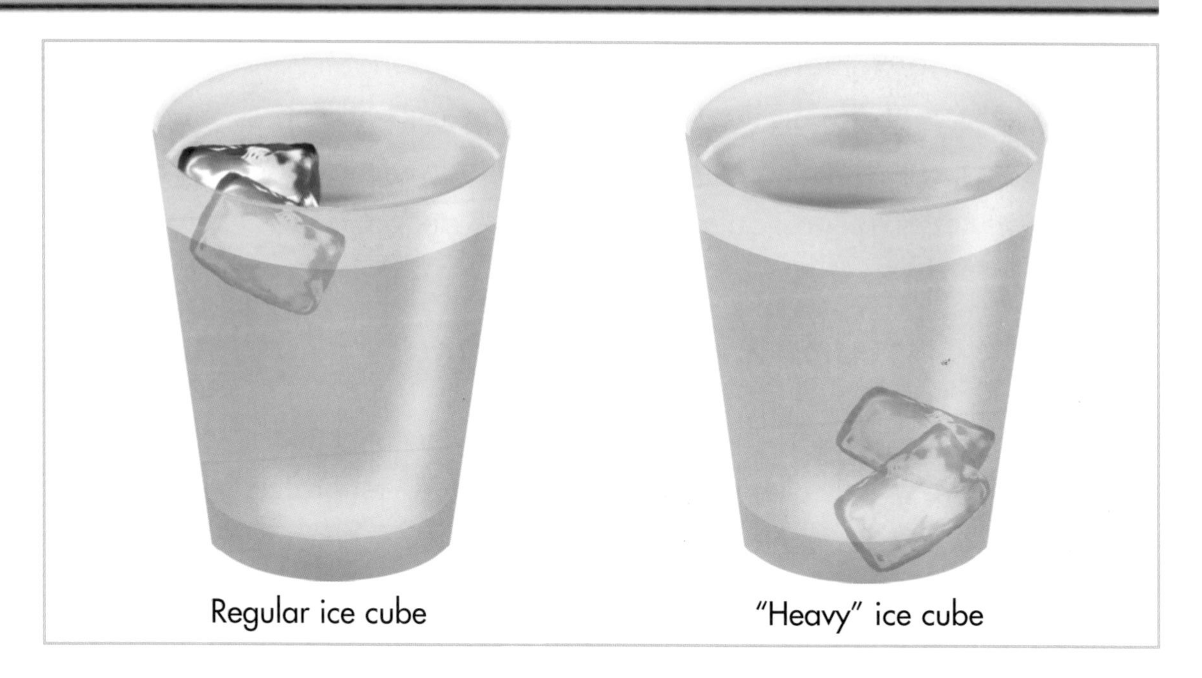

Regular ice cube

"Heavy" ice cube

How can you tell heavy water from regular water? You can't see the deuterium isotopes. But if you freeze heavy water into ice cubes, something happens that you can see. Ice cubes made from ordinary water float in water because water ice expands just a little as it freezes. But ice cubes made from heavy water will sink in ordinary water. They're just a little denser than regular water.

Hydrogen Isotopes

Deuterium is an isotope of hydrogen with a neutron added to the atom's nucleus. Tritium is a hydrogen isotope with two neutrons. These are the only isotopes that have names, as if they were elements themselves. The nucleus of hydrogen is so small and light—it has only one proton—that adding one or two neutrons changes some of its physical properties. Most of the difference is noticed at extremely low temperatures near absolute zero (the lowest possible temperature, equal to –273.15°C, where all particle motion would stop completely), where hydrogen is liquid. At ordinary temperatures humans experience every day, heavy water can absorb more heat than ordinary water.

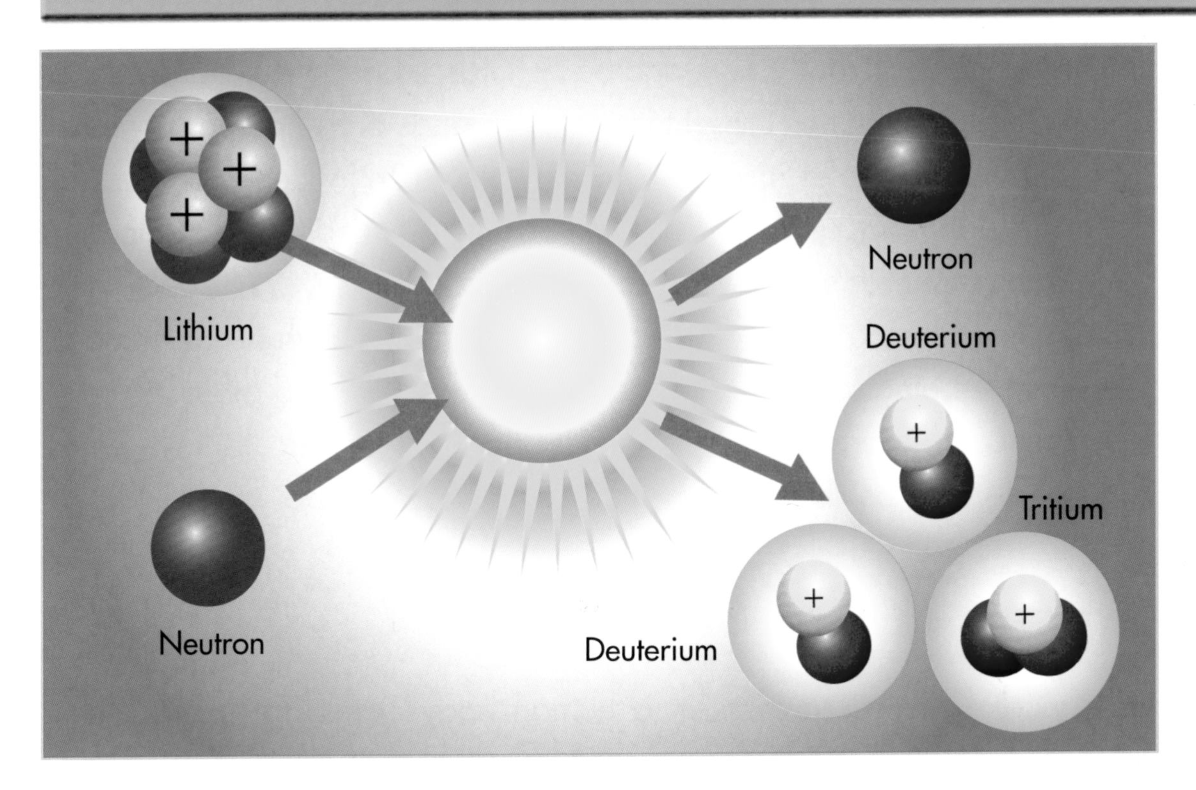

When a fast-moving neutron hits an atom of lithium, it can break the nucleus apart. Breaking the bond that holds the protons and neutrons together releases a lot of energy. One collision after another can keep happening, releasing more and more energy. For a few moments, these breaking and re-forming nuclei are glowing with the same kind of energy that makes the sun shine.

struck by fast-moving neutrons, the lithium atoms break apart to form tritium and deuterium. An atom of tritium colliding with another of deuterium will combine in a fusion reaction to produce an atom of helium and another fast-moving neutron. It will also release energy.

Lithium in Living Things

Traces of lithium are found in plants, plankton (one-celled aquatic organisms), and invertebrates (animals with no backbones, such as insects or clams). For every billion atoms in these living things, only 69 to 5,760 atoms are lithium. Vertebrates (animals with backbones) have traces of lithium in nearly all their tissue and body fluids, from 21 to 763 parts per billion. Though lithium is present in so many kinds of living things, we still do not know what roles it serves. It certainly has an influence on enzyme activity, metabolism, and respiration. It has been linked to birth defects in developing embryos when the cells are differentiating.

Lithium chloride has an effect on crystal shape when the mineral aragonite is being formed. The shells of many clams and other mollusks naturally contain aragonite. The variety of shapes and colors we enjoy in seashells may be due to the actions of lithium chloride.

Lithium in Our Bodies

How much lithium is present in a human body? Well, for every billion atoms, only twenty-seven atoms are lithium. A human who has a mass of 75 kilograms, or weighs 165 pounds, would have only 0.0002 ounces (0.007 grams) of lithium. All the lithium in twenty big people might be enough to make one tiny bead. That isn't much, compared to other elements found in the human body, such as carbon and nitrogen. But it's the right amount for our bodies. No one needs to worry about getting enough lithium. Eating an ordinary variety of plants provides all we need.

Lithium as a Prescription Medicine

The tiny amount of lithium in plants, animals, and humans must do some subtle, vital work. Exactly what is still a mystery to doctors.

The use of lithium salts was begun by Australian psychiatrist John Cade (1912–1980) after an accidental discovery of their effects on animals led to trials on human patients. Doses of lithium carbonate (Li_2CO_3), lithium citrate ($Li_3C_6H_5O_7$), or lithium orotate (LiOr) can help calm strong mood swings suffered by people with manic-depressive syndrome (also called bipolar disorder). The lithium is the active part. It relieves both depression and mania.

It is useful for some psychoses as well, but it is not a tranquilizer. Unfortunately, lithium and its compounds are also toxic, so overdoses must be avoided.

No one is sure exactly how lithium helps people with mental illnesses or mood disorders. Lithium has no known purpose in the body or in the functions of the brain. Perhaps lithium substitutes somehow for sodium or potassium. These elements have some chemical properties similar to lithium's, and they are essential for several vital processes.

Lithium in Water

Most drinking water has dissolved mineral salts. Many brands of bottled and filtered water contain naturally occurring lithium salts and other minerals.

Could drinking water with dissolved lithium have any effect on people? Some scientists believe the effect could be similar to a very mild dose of lithium carbonate. Perhaps this could help groups of people be a little calmer and get along with each other better.

There are also scientists who point out that the amount of lithium in drinkable water is probably 1 percent of a prescription dose, far too little to have any mood-altering effect. Millions of people still drink water with mineral salts from streams or wells, city taps, or bottles.

Lithium in Fireworks

When bits of petalite or spodumene are added to a fire, the flame turns red. This red flame was the first way people noticed lithium and its compounds, long before the element itself was isolated. Fireworks were invented in China hundreds of years ago. These bright sparks flying into the air can be many colors. When the sparks are red, there is lithium carbonate mixed into the gunpowder. Fireworks are a cheerful and highly visible use for this element that is hidden in stockpiles of nuclear weapons but also widespread as a few invisible atoms in our saucepans and airplanes.

The Periodic Table of Elements

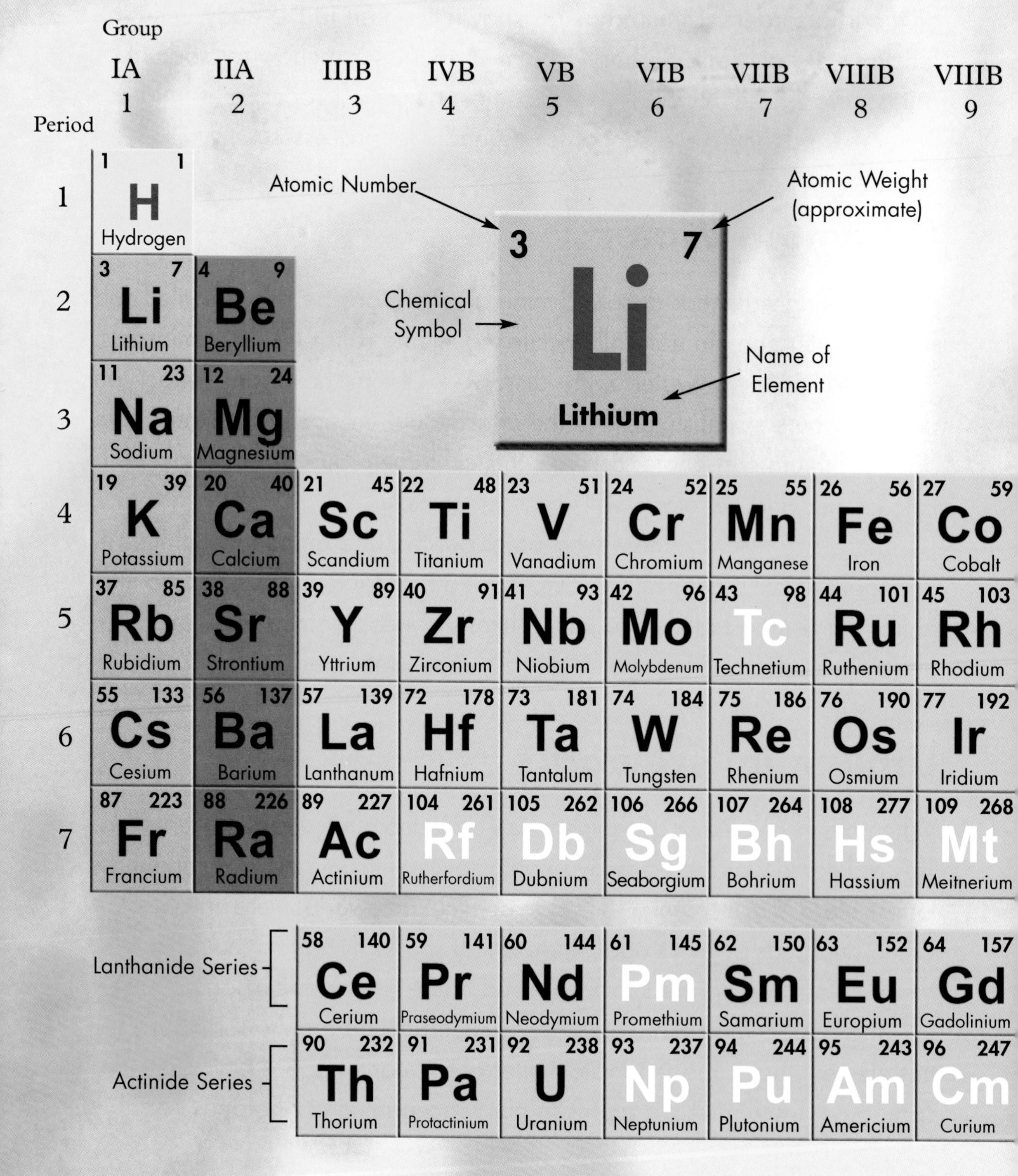

Atomic Number

Atomic Weight (approximate)

Chemical Symbol

Name of Element

3 7 Li Lithium

Period	Group IA 1	IIA 2	IIIB 3	IVB 4	VB 5	VIB 6	VIIB 7	VIIIB 8	VIIIB 9
1	1 1 H Hydrogen								
2	3 7 Li Lithium	4 9 Be Beryllium							
3	11 23 Na Sodium	12 24 Mg Magnesium							
4	19 39 K Potassium	20 40 Ca Calcium	21 45 Sc Scandium	22 48 Ti Titanium	23 51 V Vanadium	24 52 Cr Chromium	25 55 Mn Manganese	26 56 Fe Iron	27 59 Co Cobalt
5	37 85 Rb Rubidium	38 88 Sr Strontium	39 89 Y Yttrium	40 91 Zr Zirconium	41 93 Nb Niobium	42 96 Mo Molybdenum	43 98 Tc Technetium	44 101 Ru Ruthenium	45 103 Rh Rhodium
6	55 133 Cs Cesium	56 137 Ba Barium	57 139 La Lanthanum	72 178 Hf Hafnium	73 181 Ta Tantalum	74 184 W Tungsten	75 186 Re Rhenium	76 190 Os Osmium	77 192 Ir Iridium
7	87 223 Fr Francium	88 226 Ra Radium	89 227 Ac Actinium	104 261 Rf Rutherfordium	105 262 Db Dubnium	106 266 Sg Seaborgium	107 264 Bh Bohrium	108 277 Hs Hassium	109 268 Mt Meitnerium

Series							
Lanthanide Series	58 140 Ce Cerium	59 141 Pr Praseodymium	60 144 Nd Neodymium	61 145 Pm Promethium	62 150 Sm Samarium	63 152 Eu Europium	64 157 Gd Gadolinium
Actinide Series	90 232 Th Thorium	91 231 Pa Protactinium	92 238 U Uranium	93 237 Np Neptunium	94 244 Pu Plutonium	95 243 Am Americium	96 247 Cm Curium

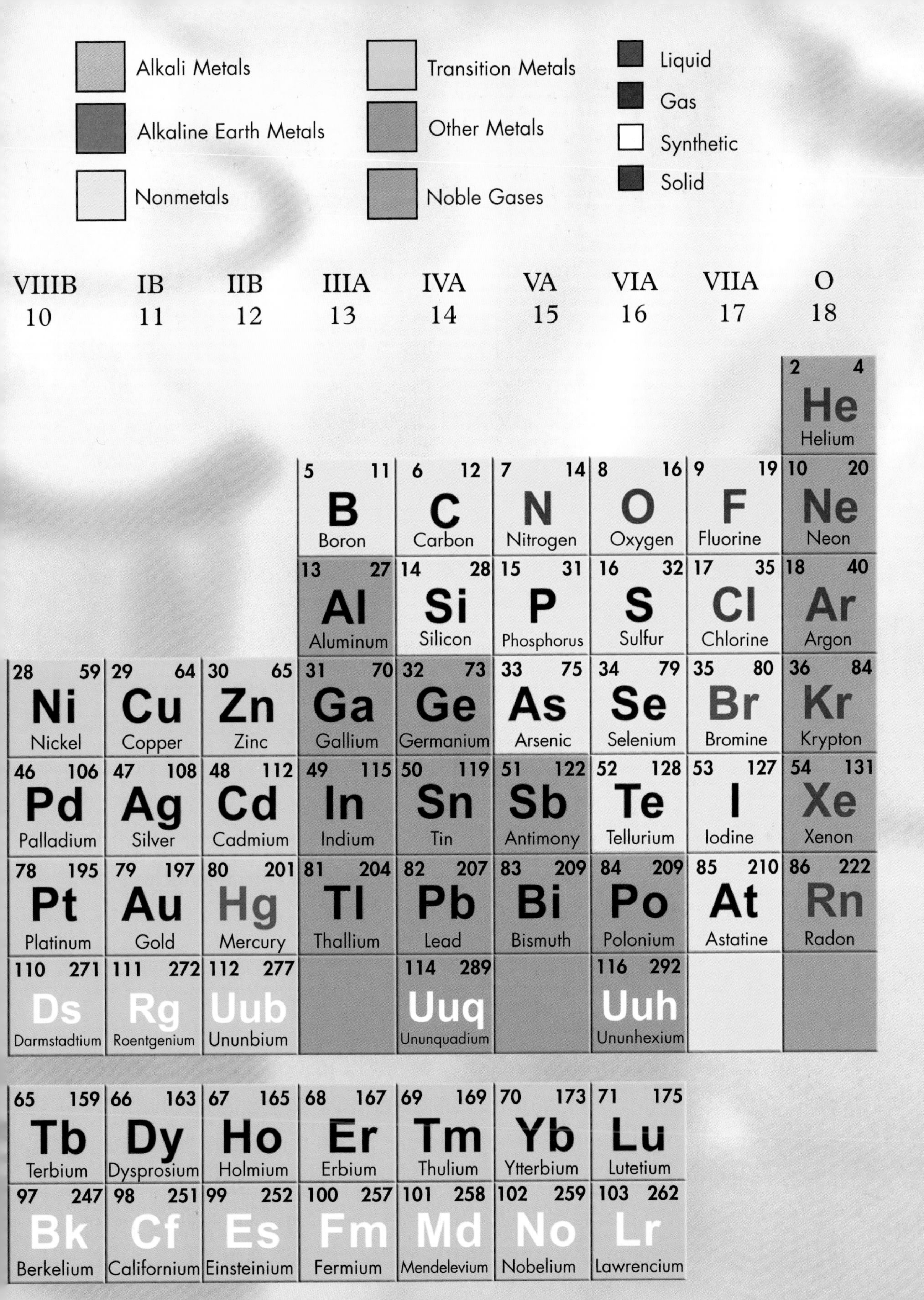
Alkali Metals
Alkaline Earth Metals
Nonmetals
Transition Metals
Other Metals
Noble Gases
Liquid
Gas
Synthetic
Solid
VIIIB 10
IB 11
IIB 12
IIIA 13
IVA 14
VA 15
VIA 16
VIIA 17
O 18
2 4 He Helium
5 11 B Boron
6 12 C Carbon
7 14 N Nitrogen
8 16 O Oxygen
9 19 F Fluorine
10 20 Ne Neon
13 27 Al Aluminum
14 28 Si Silicon
15 31 P Phosphorus
16 32 S Sulfur
17 35 Cl Chlorine
18 40 Ar Argon
28 59 Ni Nickel
29 64 Cu Copper
30 65 Zn Zinc
31 70 Ga Gallium
32 73 Ge Germanium
33 75 As Arsenic
34 79 Se Selenium
35 80 Br Bromine
36 84 Kr Krypton
46 106 Pd Palladium
47 108 Ag Silver
48 112 Cd Cadmium
49 115 In Indium
50 119 Sn Tin
51 122 Sb Antimony
52 128 Te Tellurium
53 127 I Iodine
54 131 Xe Xenon
78 195 Pt Platinum
79 197 Au Gold
80 201 Hg Mercury
81 204 Tl Thallium
82 207 Pb Lead
83 209 Bi Bismuth
84 209 Po Polonium
85 210 At Astatine
86 222 Rn Radon
110 271 Ds Darmstadtium
111 272 Rg Roentgenium
112 277 Uub Ununbium
114 289 Uuq Ununquadium
116 292 Uuh Ununhexium
65 159 Tb Terbium
66 163 Dy Dysprosium
67 165 Ho Holmium
68 167 Er Erbium
69 169 Tm Thulium
70 173 Yb Ytterbium
71 175 Lu Lutetium
97 247 Bk Berkelium
98 251 Cf Californium
99 252 Es Einsteinium
100 257 Fm Fermium
101 258 Md Mendelevium
102 259 No Nobelium
103 262 Lr Lawrencium

Glossary

alkali metals Elements from group 1 in the periodic table, which are very reactive.

atomic number The number of protons in the nucleus of an element.

atomic weight Also known as atomic mass. The average of the weights or masses of all the naturally occurring isotopes of a specific element.

compound A substance that is made from the atoms of two or more elements joined together by chemical bonds.

density How much mass an object contains in a given volume (mass per unit volume) which is often expressed in grams per cubic centimeter (g/cm^3).

heavy water A molecule of water in which one or both atoms of hydrogen have been replaced with deuterium or tritium, isotopes of hydrogen.

ion An atom or molecule with an electric charge due to the loss or gain of electrons.

isotopes Atoms of a specific element with different numbers of neutrons.

mass number The number of protons and neutrons in the nucleus of an atom of an element.

petalite An ore that contains lithium, aluminum, and silica.

specific heat The amount of heat needed to raise the temperature of one gram of a substance by 1°C.

transition metals The elements in groups 3 through 12 of the periodic table. They are ductile and malleable, and they conduct heat and electricity. Their valence electrons are present in more than one shell, which means that they can form more than one positive ion.

valence electrons Electrons in the outer shell of atoms; they allow atoms to link together chemically and metals to conduct heat and electricity.

For More Information

Journal of Chemical Education
University of Wisconsin-Madison
209 N. Brooks Street
Madison, WI 53715-1116
(800) 991-5534
Web site: http://jchemed.chem.wisc.edu/index.html

Oregon Museum of Science and Industry
1945 Southeast Water Avenue
Portland, OR 97214-3354
(503) 797-4000
Web site: http://www.omsi.edu

Royal British Columbia Museum
675 Belleville Street
Victoria, BC V8W 9W2
Canada
(250) 387-2478
Web site: http://www.royalbcmuseum.bc.ca

Web Sites

Due to the changing nature of Internet links, Rosen Publishing has developed an online list of Web sites related to the subject of this book. This site is updated regularly. Please use this link to access the list:

http://www.rosenlinks.com/uept/lith

For Further Reading

Gonick, Larry, and Craig Criddle. *The Cartoon Guide to Chemistry.* New York, NY: HarperCollins, 2005.

Hudson, John. *The History of Chemistry.* New York, NY: Routledge, 1992.

Knapp, Brian J. *The Periodic Table.* Danbury, CT: Grolier Educational, 1998.

Stwertka, Albert. *A Guide to the Elements.* 2nd ed. New York, NY: Oxford University Press, 2002.

Bibliography

Cotton, F. Albert, Geoffrey Wilkinson, Carlos A. Murillo, and Manfred Bochmann. *Advanced Inorganic Chemistry.* 6th ed. New York, NY: John Wiley & Sons, 1999.

Gray, Theodore. "The Wooden Periodic Table." Retrieved February 3, 2006 (http://www.theodoregray.com/PeriodicTable).

Stwertka, Albert. *A Guide to the Elements.* 2nd ed. New York, NY: Oxford University Press, 2002.

Tweed, Matt. *Essential Elements: Atoms, Quarks and the Periodic Table.* New York, NY: Walker & Company, 2003.

Index

About the Author

Paula Johanson has worked for twenty years as a writer and teacher, and she has written and edited curriculum educational materials for the Alberta Distance Learning Centre in Canada. She writes and edits nonfiction books, magazine articles and columns, and book reviews. At two or more conferences each year, she leads panel discussions on practical science (usually biochemistry) and how it applies to home life and creative work. Her pottery and clay figures are glazed with lithium salts. Ms. Johanson lives on an island in British Columbia and on a farm in Alberta, Canada.

Photo Credits

Cover, pp. 1, 7, 10, 12, 14, 29, 38, 39, 42–43 by Tahara Anderson; p. 5 © Alan Schein/zefa/Corbis; pp. 17, 23 © Andrew Lambert Photography/Photo Researchers, Inc.; p. 21 courtesy LPS ® Laboratories, a division of Illinois Tool Works; p. 25 courtesy of FMC Corporation, Lithium Division; p. 26 © Matthias Kulka/Corbis; pp. 27, 33 by Mark Golebiowski; p. 35 (left) © Stu Griffith/ iStockphoto; p. 35 (right) © iStockphoto; p. 36 © NASA.

Special thanks to Jenny Ingber, high school chemistry teacher, Region 9 schools, New York City, NY, for her assistance in executing the science experiments illustrated in this book.

Designer: Tahara Anderson; Editor: Kathy Kuhtz Campbell